Newport Community Learning
and Libraries
Z811089

D1631832

DK eyewonder

Earth

LONDON, NEW YORK,
MELBOURNE, MUNICH, and DELHI

Written and edited by Penelope York
Designed by Cheryl Telfer and Helen Melville
Managing editor Susan Leonard
Managing art editor Cathy Chesson
Jacket design Chris Drew
Picture researcher Marie Osborn
Production Shivani Pandey
DTP designer Almudena Díaz
Consultant Chris Pellant

REVISED EDITION
DK UK
Senior editor Caroline Stamps
Senior art editor Rachael Grady
Jacket editor Manisha Majithia
Jacket designer Natasha Rees
Jacket design development manager
Sophia M Tampakopoulos Turner
Producer (print production) Mary Slater
Producers (pre-production)
Francesca Wardell, Rachel Ng
Publisher Andrew Macintyre

DK INDIA
Senior editor Shatarupa Chaudhuri
Senior art editor Rajnish Kashyap
Editor Surbhi Nayyar Kapoor
Art editor Amit Verma
Managing editor Alka Thakur Hazarika
Managing art editor Romi Chakraborty
DTP designer Dheeraj Singh
Picture researcher Sumedha Chopra

First published in Great Britain in 2002
This edition published in Great Britain in 2015
by Dorling Kindersley Limited
80 Strand, London WC2R 0RL

13 14 15 16 17 10 9 8 7 6 5 4 3 2 1
001 – 196174 – 02/15

A CIP catalogue record for this book
is available from the British Library.

ISBN 978-1-4093-3698-3

Colour reproduction by Scanhouse, Malaysia
Printed and bound in China by Hung Hing

Discover more at
www.dk.com

Contents

Where are we?

Where is Earth? Good question. Let's look into space and find out where we are and what is around us. Then we'll zoom in closer.

Mercury

Venus

Earth

Mars

Sun

Let's zoom in on Earth.

Can you see the towns?

Earth from space

When we zoom in and take a look at our Earth from space, we can see how the countries and oceans are laid out. You are somewhere down there. This is a photograph of the USA taken by a satellite.

Spotting cities

When we look a bit closer we start to see built-up city areas and green country areas. You are now looking at Florida, a state in the USA. Can you see anyone yet?

The Solar System

Our Earth is in the middle of a family of planets that all move around our Sun. We call this the Solar System. So far, life has not been discovered on any other planet besides Earth, but it soon might be one day.

Hunting down houses

Diving down a bit, we can now see a town in Florida next door to the beach. But we still can't see any people down there.

Finding people

Zoom in on a house and at last, we can see kids! Now look back at Earth and you'll soon realize how big it is. It's absolutely enormous.

Crust to core

We think we know so much about Earth and even about space, but what lies beneath our feet? Imagine that Earth is an apple. The crust that we stand on would be as thick as the apple skin. That leaves a lot of something else underneath.

Journey to the centre of the Earth

Man has only dug about 13 km (8 miles) into Earth, which is only about five-hundreth of the journey to the centre. Scientists can only guess what is beneath but we do know that it is very, very hot.

Earth facts

- You may think Earth is big, but the Sun could swallow up 1,303,600 Earths.
- If you wanted to walk all the way around Earth along the equator, then it would take you about a whole year, non-stop. You wouldn't even be able to sleep!

All around Earth is a blanket called the atmosphere, which contains the air we breathe.

The crust is the thin layer of rock that covers Earth. It can be between 5 and 68 km (3½ and 42 miles) thick.

Granite

Basalt

Peridotite

Earth's surface

Earth is made up of rocks. Granite is a typical continental (land) rock. Basalt is a typical ocean floor rock, and peridotite is a mantle rock.

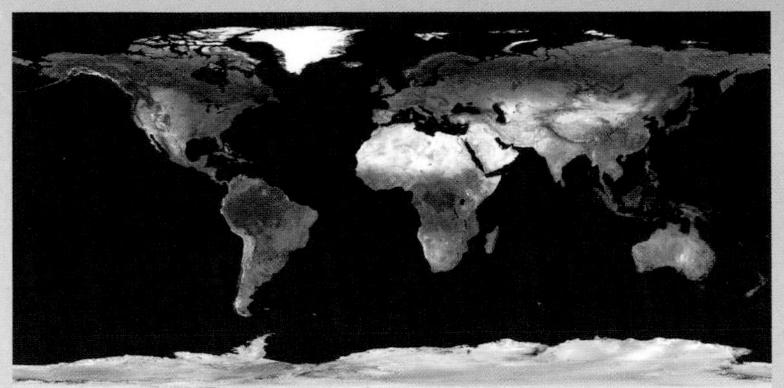

Earth map

About 29% of Earth's surface is made up of land, which is divided into seven continents (a piece of land that is not broken up by sea). These are North America, South America, Europe, Africa, Australasia, Asia, and Antarctica.

People only live on 12% of Earth's surface.

The mantle is the layer that lies below the crust. The deeper mantle is solid rock but the upper layers are plastic, moving rock.

The outer core is completely liquid. It is made of iron and nickel.

The inner core is a red hot, solid ball of molten iron and nickel that is 4,500°C (8,132°F). That's hot!

Moving world

Earth's crust is made up of huge plates, which fit together like a jigsaw. The plates have been moving for millions of years and still shift today, with dramatic effects on the shape of our planet's surface.

The continents ride slowly on plates of crust.

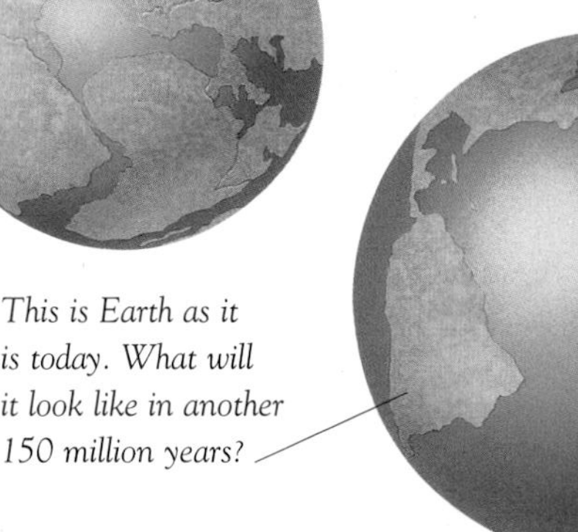

This is what the continents looked like 200 million years ago.

The continents we know today started to take shape 150 million years ago.

This is Earth as it is today. What will it look like in another 150 million years?

Plate line

A fault is a line along which two plates run side by side. When the plates move against each other, they can create earthquakes, volcanoes, or even mountains.

Slow progress

The plates drift in certain directions. As they shift, they change in shape and size – this takes many millions of years. See what Earth looked like 200 million years ago compared to today.

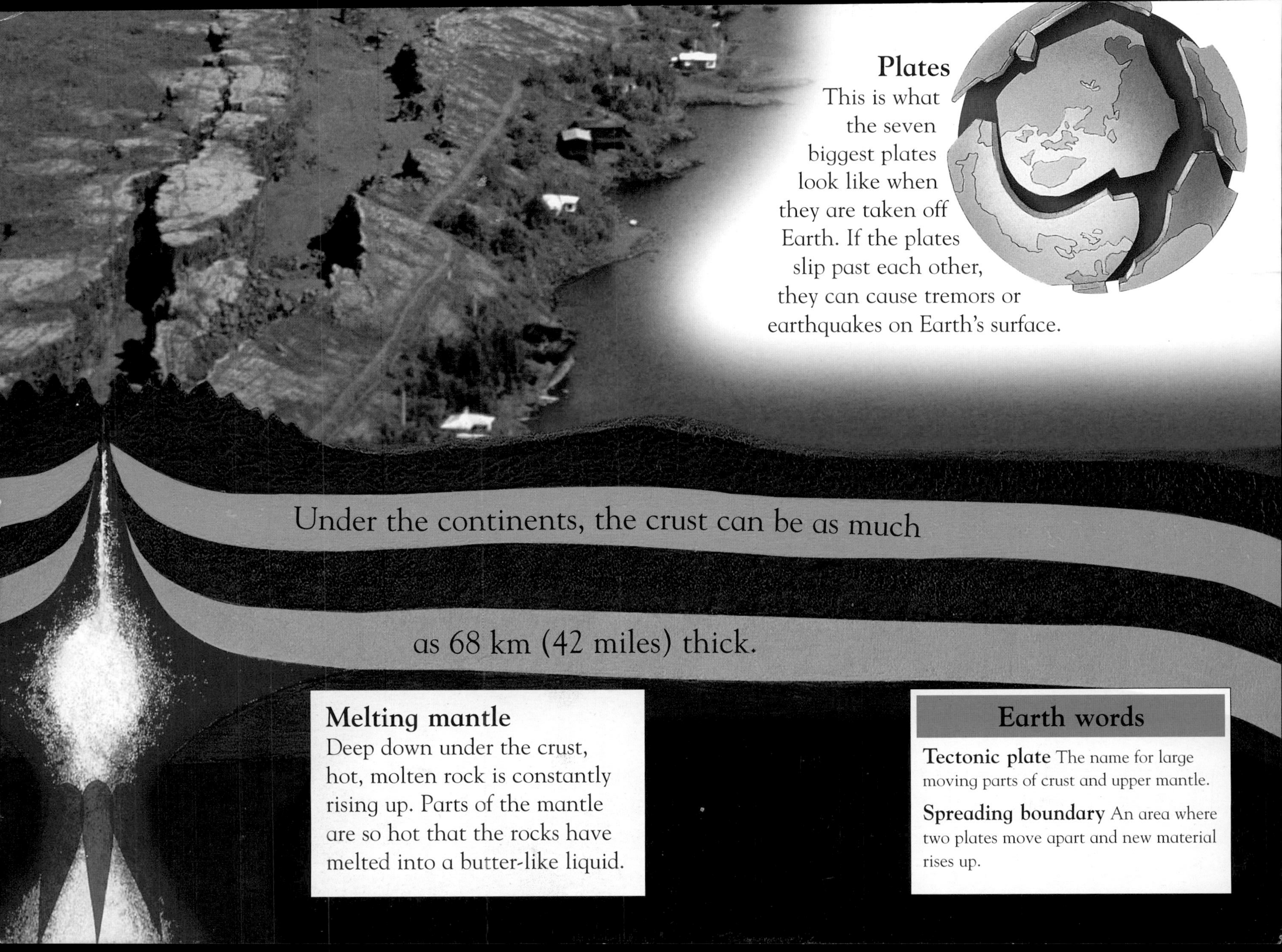

Plates

This is what the seven biggest plates look like when they are taken off Earth. If the plates slip past each other, they can cause tremors or earthquakes on Earth's surface.

Under the continents, the crust can be as much as 68 km (42 miles) thick.

Melting mantle

Deep down under the crust, hot, molten rock is constantly rising up. Parts of the mantle are so hot that the rocks have melted into a butter-like liquid.

Earth words

Tectonic plate The name for large moving parts of crust and upper mantle.

Spreading boundary An area where two plates move apart and new material rises up.

The tips of the world

Without mountains, Earth would look far less spectacular. About 5% of the world's land surface is made up of impressive highland.

Old mountains

Mountains are made when Earth's crust is pushed up in big folds or forced up or down in blocks. The 450-million-year-old Scottish Highlands used to be craggy like the Himalayas (below), but wind and rain have worn them down.

New mountains

The Himalayas, in Asia, are good examples of fold mountains. They are 50 million years old, which is relatively new! Mount Everest in the Himalayas is the highest point on Earth.

The plate pushes forwards slowly over the years, making more and more folds.

This model shows how plates push together, from the left side, forcing one side to crumple into mountains.

Block mountains

Block mountains are formed when Earth's crust is moved up or down in blocks. Mount Rundle, Banff National Park, Canada, is a spectacular example of a block mountain.

Fault lines occur and a block drops or lifts to produce a high mountain and a low plain.

Hawaii is the tip of a very, very big mountain.

Underwater mountains

Long lines of islands in the oceans are actually the tips of huge mountain ranges, which lie underwater. Mauna Kea, on the island of Hawaii, is the world's tallest mountain from the bottom of the sea to the tip.

The Himalayas are still rising by 4 mm ($\frac{1}{6}$ in) every year.

The Himalayas began to form when India collided with Asia.

The fire mountain

Pressure builds up underground. Hot, liquid rock, called magma, finds its way to a weak part between Earth's plates and explodes. Welcome to the volcano.

The big killer

The force of an exploding volcano is enormous – like opening a can of shaken, fizzy drink. Chunks of molten rock as big as houses can be flung high into the air and dust can travel as much as 20 km (13 miles) high.

Mountain makers

As the insides of Earth explode out of the ground, the lava and ash settle and over time a perfectly shaped mountain is formed. In effect, Earth is turning a little bit of itself inside out.

The lava that bursts out of a volcano is 10 times hotter than boiling water in a kettle.

Rivers of fire

When magma pours out of volcanoes it is called lava. It rolls slowly downhill in a huge river, burning everything in its path. When it cools, it solidifies into rock, called igneous rock.

Bubble trouble

In some volcanic areas, you can see heat coming up from under the ground. Mud bubbles and hot water jets, called geysers, shoot up high. They sometimes smell of rotten eggs because of a gas called hydrogen sulphide.

KILLER GAS

Sometimes the gas that comes out of a volcano is poisonous. In 79 CE, Mount Vesuvius, Italy, erupted violently. A cloud of gas rolled down and poisoned many people in Pompeii, the town at its base. Ash buried them and casts have been made from the spaces the bodies left.

Earthquake!

Imagine waking up one night to find the ground trembling and shaking. That's what it's like to feel an earthquake. These sudden movements in Earth's plates can cause terrifying damage.

Fault line

The deadly tsunami

When an earthquake happens underwater, vibrations cause ripples in the sea. They grow and grow until they are enormous, deadly waves, or tsunami, that crash onto the shore.

Whose fault?

An earthquake is caused when two of Earth's plates slide against each other. The line that they slide along is called a fault. When they move they cause vibrations across the ground.

Shock waves caused by an earthquake are recorded by a machine called a seismometer.

Devastation

Earthquakes can be so strong that they cause whole buildings to collapse. Children who live in areas that have earthquakes are trained regularly on how to remain safe.

The most powerful earthquakes are in Japan.

One in 1923 killed 143,000 people.

The rock cycle

Geologists divide the rocks that make up Earth's crust into three groups: igneous, sedimentary, and metamorphic. But they all come from the same original material, which moves round in a big cycle.

Igneous rock

Granite and basalt are typical examples of igneous rock. They start their lives as melted rock, such as underground magma and lava, which comes out of volcanoes.

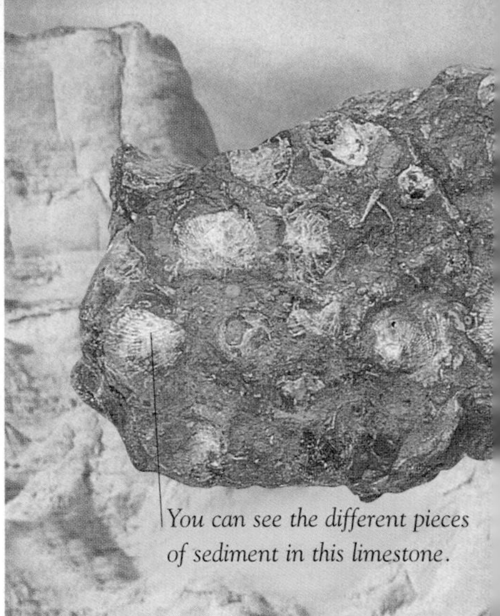

You can see the different pieces of sediment in this limestone.

Chalk is also a type

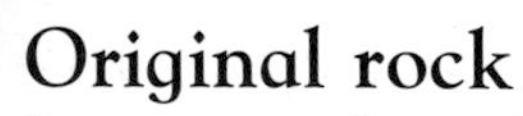

Original rock

Igneous rock cools down and hardens either beneath the surface or on the surface when it erupts from a volcano. It is rock from deep in Earth's crust.

Break down

Little pieces of igneous rock are broken off by rain and wind and are carried to the sea where they pile up as layers of sediment. Th remains of sea creatur are buried in the layer and may become fossi

Sedimentary rock

Limestone and sandstone are typical sedimentary rocks. You can often see the layers in a big piece of rock when they have been squashed down on the sea bed, such as in this photo of the Painted Desert in Arizona, USA.

Metamorphic rock

Marble is a good example of this rock. It was limestone before it was changed by intense heat. When marble is polished it produces beautiful swirly patterns.

This man is chiselling a square piece of marble from a marble quarry in Italy.

Under pressure

When sedimentary or igneous rocks are dragged deep into Earth's crust by the moving plates, they can get hot, changing into a new rock type – metamorphic. If the rocks are taken so deep that they melt, they are back where they started as magma. The cycle starts again.

Vital survival

All around Earth is a protective shield called the atmosphere. It keeps us from burning under the Sun during the day and from freezing at night. Within our atmosphere lie the water and air cycles.

The water cycle

It's incredible to imagine, but the water that we use every day is the same water as was on Earth millions of years ago. It goes up into the clouds, and back down to Earth as rain, and never stops its cycle.

Water goes up and

Air goes in and ou

Water, water everywhere

Water goes up and water comes down. It is evaporated into the atmosphere by the Sun and turns into clouds. When the clouds cool down high up in the sky, rain falls from them.

down, up and down.
nd around and around.

The air cycle

The air that we breathe is also in a continuous cycle. Animals breathe in a gas called oxygen and breathe out carbon dioxide. All plants take in carbon dioxide and make oxygen.

Essential air

Because of the air cycle between animals and plants, we could not possibly live without each other. We make the air for each other that is vital for life.

Down to earth

Without soil, life would be impossible as nothing can grow without it. Soil is the part of Earth that lies between us and the solid bedrock.

Out of the soil grow many plants.

This level is called topsoil. It is rich in food for plants and contains living creatures.

The subsoil has less nutrients for plants to feed on.

As you get lower, the soil becomes rockier.

The solid rock below the soil is called bedrock.

Useful soil

Soil can be used in so many ways, from making bricks to providing clay for pottery, but it is most vital for growing plants for us to eat. In southeast Asia, people build hillside terraces to stop soil from washing away when it rains.

Layers of soil

If you cut a section through the soil, down to the rock beneath, you would find lots of layers. The material nearest the top is the rich soil needed for plants to grow and the bottom is solid rock.

A handful of soil contains about six billion bacteria!

What is soil?

Soil is made up of rocks, minerals, dead plants and animals, tiny creatures, gases, and water. As plants and animals die, tiny creatures and bacteria break them down to become soil.

Essential food

Plant roots take in water and nutrients from the soil. Plants need these in order to make food and grow.

Wriggling worms

Worms are vital to the soil. They eat decaying plants and animals and deposit them into the soil as they wriggle through it. As they tunnel, they help the soil to breathe.

Nature's sculptures

The moment rock is exposed to air, it is in serious danger of being reshaped. Wind, rain, air, and even plants, all seem determined to change the landscape, sometimes with beautiful results.

Frost shattering

This amazing landscape was carved by frost. Water in cracks in the rock freezes and expands and the rock shatters, leaving extraordinary, spiky shapes.

Pillars of Earth

These strange pillars are called hoodoos. They are formed because soft rock lies below hard rock. Downpours of rain wash away the softer rock, leaving pillars of harder rock above.

Limestone pavement

Limestone is a soft rock that is affected dramatically by rainwater. The slightly acid rainwater changes the limestone into a softer rock, which is washed away. Cracks get larger and the ground becomes uneven.

When air, wind, ice, or plants change the shapes of rock, it is called "weathering".

Watch out! Plant attack

Trees sometimes speed up rock cracking with their roots. As the roots grow, they creep between cracks; when they thicken, they force the cracks to open wider.

Flow of water

Water is incredibly powerful. When there is a lot of it, moving at huge speeds, it can carry away a lot of loose rock and mud. When water changes the shape of a landscape, it is called erosion.

River facts

- The longest river in the world is the Nile in Egypt. Beneath the Nile runs another river deep underground that holds six times as much water.
- The highest waterfall in the world is Angel Falls in Venezuela. The water falls 3¾ times the height of the Eiffel Tower in Paris.

Running wild

As water rushes from its source, in the highlands, down to the sea, it constantly picks up chunks of rock, sand, and mud along the way. It then deposits these elsewhere, changing the shape of the land as it goes.

This harder rock is left behind after floods.

Desert floods

Water can even shape the desert. Heavy floods sometimes rush through the land, taking the land with it and leaving weird towers of rock behind, such as in Monument Valley, Arizona, USA.

Water power

The Grand Canyon, USA, is the largest gorge in the world. It has been carved by the Colorado River over 20 million years. Different rocks react in different ways to the water, so the shapes are incredibly spectacular

Niagara Falls

Niagara Falls is an enormous waterfall, which moves backwards by 1 m (3 ft 3 in) each year. The lower rock is soft and is worn away by the water. Eventually, the top rock crashes to the bottom when it can no longer stay where it is without support.

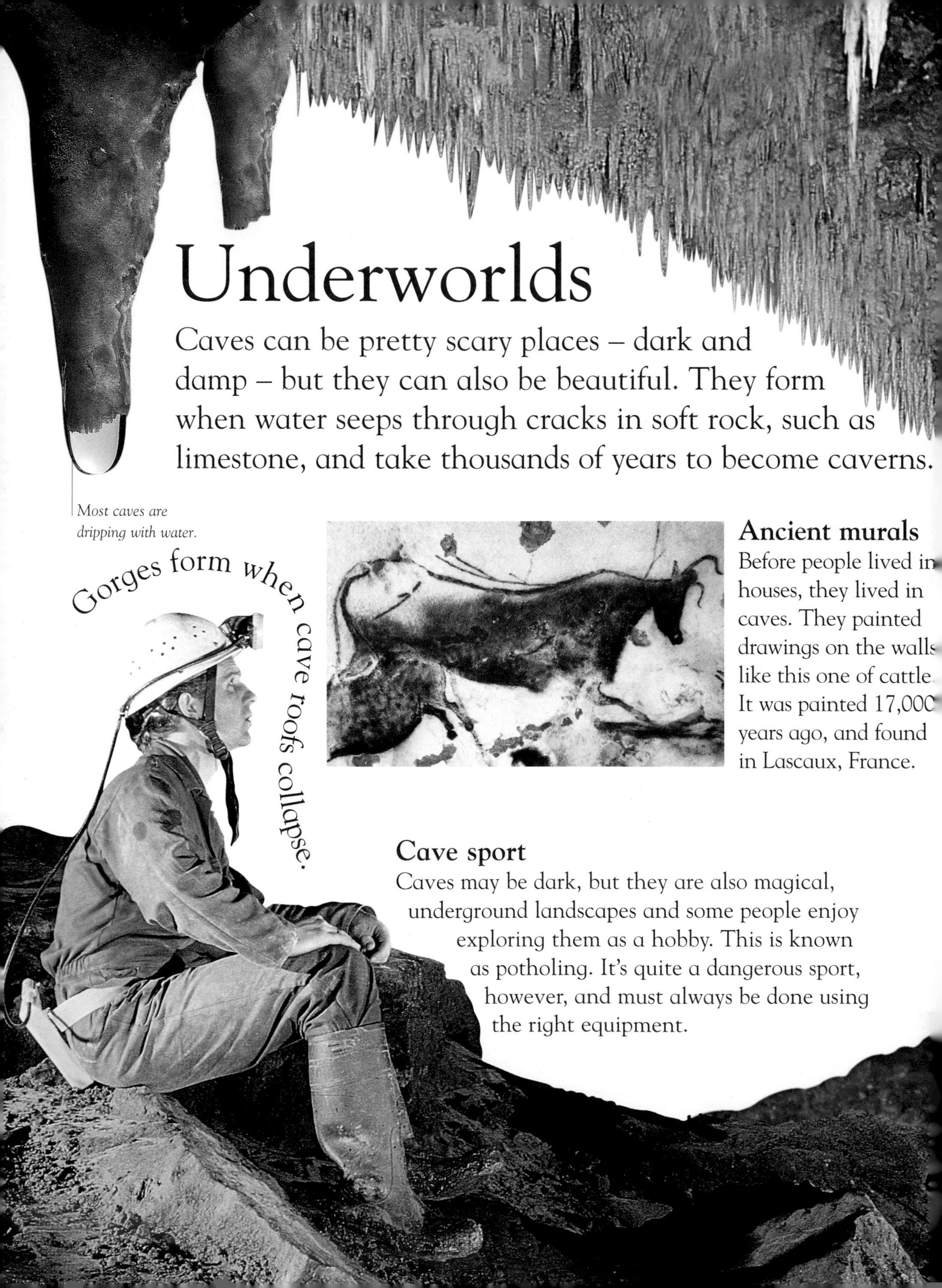

Underworlds

Caves can be pretty scary places – dark and damp – but they can also be beautiful. They form when water seeps through cracks in soft rock, such as limestone, and take thousands of years to become caverns.

Most caves are dripping with water.

Gorges form when cave roofs collapse.

Ancient murals

Before people lived in houses, they lived in caves. They painted drawings on the walls like this one of cattle. It was painted 17,000 years ago, and found in Lascaux, France.

Cave sport

Caves may be dark, but they are also magical, underground landscapes and some people enjoy exploring them as a hobby. This is known as potholing. It's quite a dangerous sport, however, and must always be done using the right equipment.

The biggest cave in the world, the Son Doong Cave, Vietnam, could fit hundreds of tennis courts inside it.

It can take thousands of years for stalactites and stalagmites to meet.

Up and down

When water drips from a cave's ceiling, it leaves behind tiny pieces of rock. Over many years, the pieces grow downwards into a long icicle shape called a stalactite. Where the water falls to the ground, stalagmites grow upwards like candles.

The power of ice

There's more to snow and ice than meets the eye. Not only do they produce some of the most spectacular scenes on Earth, but they also are powerful tools that sculpt it.

Earth's natural plough

A glacier is an enormous mass of ice that flows downhill slowly. When glaciers melt, they show how much of Earth has been gorged away. You can see how a glacier has shaped this Norwegian fjord.

The mighty glacier

A glacier is incredibly powerful. It carves its way through mountains, leaving huge gorges or valleys behind. On the way it swallows up and moves giant boulders. Yet it only moves at a speed of about a centimetre or two a day.

Floating island

Some icebergs are huge. But whatever you see above water, there is even more below. Two-thirds lies underwater.

Ice caves

Icebergs are big blocks of ice that break away from the end of a glacier. As they melt, the wind and waves batter them into weird shapes, sometimes creating ice caves.

The mighty wave

When you play on a sandy beach, have you ever noticed how often the waves crash onto it? Well, believe it or not, that wave movement is constantly changing the coastline. Waves are even powerful enough to reshape cliffs!

Waves destroy

Shock waves

- As the waves force coastlines back, sometimes houses built on the cliffs fall into the sea!
- A series of pounding 10-m- (33-ft-) high storm waves can remove one whole metre (3 ft 3 in) of cliff in one night.

Making a bay

The sea is very persistent. When it finds a weak part along a coast, it breaks through and spreads out as far as it can. It eventually creates a bay, such as Wineglass Bay in Tasmania.

some coasts but make brand new beaches elsewhere.

Creating sand

Sandy beaches take hundreds of years to form. Waves near the shore pummel boulders into pebbles, and with more battering they eventually become the soft, fine-grained sand that you find on a beach.

It's amazing that just water can turn this pebble into fine sand.

Coastline sculpture

This picture shows how powerful waves are. The sea has completely battered its way through the rock on this cliff and formed an archway. Eventually, when the arch gets too weak, it falls in on itself, leaving stacks behind.

The ocean floor

The ocean is a mysterious place – we can't go beyond certain depths because the pressure will kill us. However, we do know that the ocean floor has some features that are very similar to those found on land.

Coral reef

Coral reefs are found in clear, warm waters near the shore. Corals are living things and are home to hundreds of others as well.

DEPTH GAUGE

Sea level
0 km
(0 miles)

Coral from surface to 0.5 km (⅓ mile)

Earth oceans

More than two-thirds of Earth is covered in water. The deepest part of the ocean is the Mariana Trench, in the Pacific Ocean, which is 11.5 km (7 miles) deep. Very little life can survive in those depths.

Black smokers

Where the ocean plates move against each other, vents open and hot steam rises into the water. These are called

Diving down

It is very difficult for humans or submarines to go down deep underwater. This submarine is called the Nautile and can take three people down to depths of 4 km (2½ miles)

Earth is divided into seven continents.
See page 7
Animals breathe in carbon dioxide and breathe out oxygen.
See page 19
Dinosaurs appeared about 20 million years ago.
See page 38
Basalt is a typical ocean floor rock.
See page 6
When air, wind, ice, or plants change the shapes of rock, it is called "weathering".
See page 23

What's this?

Take a look at these close-ups of pictures in the book, and see if you can identify them. The clues should help you!

- They cover about 5% of the world's land surface.
- They are made when the Earth's crust is pushed up in big folds or blocks.

See page 10

- It pours out of volcanoes.
- It is 10 times hotter than boiling water in a kettle.
- On cooling, it solidifies.

See page 13

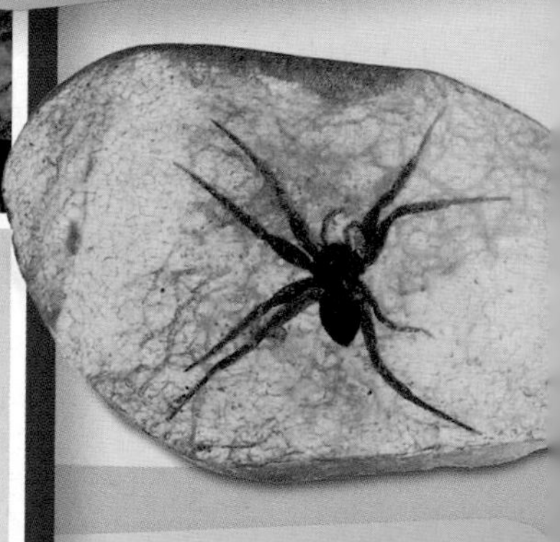

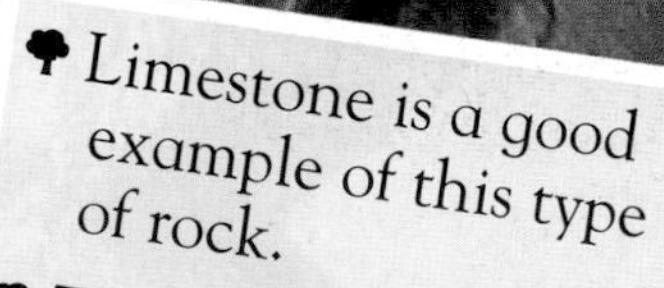

- Limestone is a good example of this type of rock.
- This type of rock has layers.

See page 17

- These can be formed by an underwater earthquake.
- They can build up to become deadly waves.

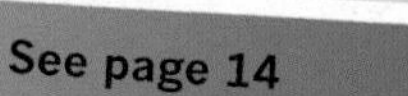

See page 14

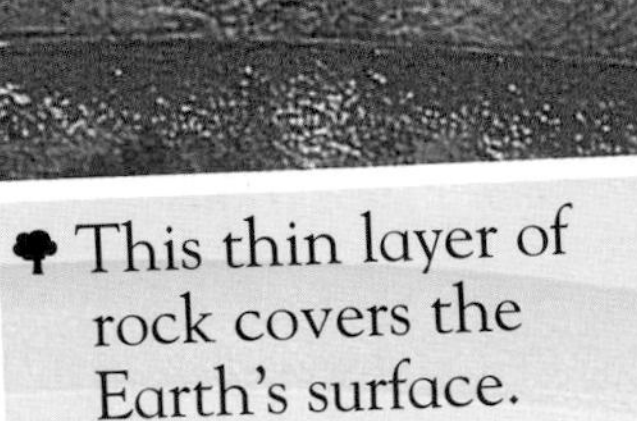

- This thin layer of rock covers the Earth's surface.
- It can be between 5 and 68 km (3½ and 42 miles) thick.

See page 6

Pollution materials
are in the wrong pla
that environment fo
and creatures that li

Rainforest a tropic
that receives heavy
therefore where hug
of plants grow.

Rock a large, solid
underground that is
exposed at the surfa
Earth and is made u
or more minerals.

Satellite an object
revolves around Ear

Sedimentary rock
in layers by the dep
eroded grains.

Seismometer an ins
measures the strengt

Sewage rubbish or
is carried away in se

Solar power energy
gained by using the

Solar System our f
eight planets that r
around our Sun.

Stack a rock pillar
in coastal waters w
of an arch falls in.

Stalactite a hangin
structure formed in
water with traces of

Stalagmite a rising
structure formed wh
drip to the floor an
of rock behind.

Strata layers of se

Tsunami a huge, f
wave that is caused
underground earth

Volcano where ho
breaks through Ear
crust with great pre

Weathering the b
of rocks by wind, r

- These are the driest places on Earth.
- They are often found in regions that get little rain, such as Africa.

See page 40

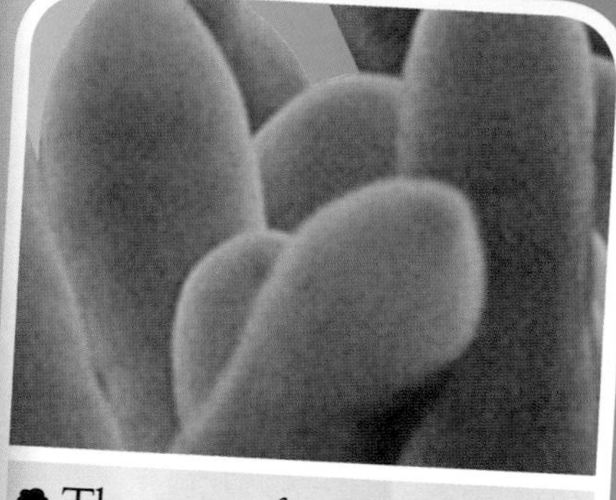

- These are found in clear, warm waters.
- They are living things and are home to other creatures as well.

See page 32

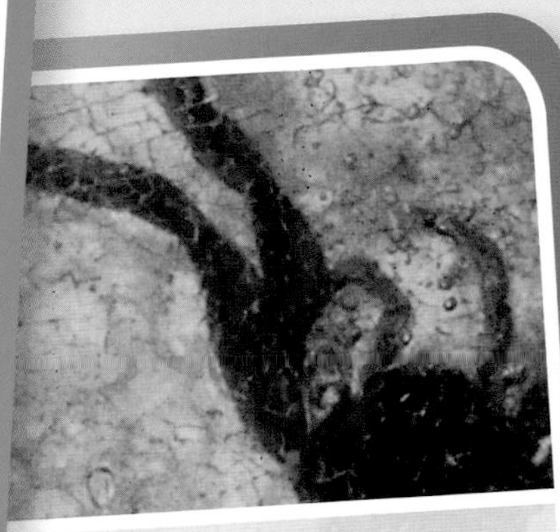

- This is fossilized tree resin or gum.
- Millions of years ago, insects got trapped in it.

See page 39

- They look similar to long icicles.
- They are formed when water drips from a cave's ceiling.

See page 27

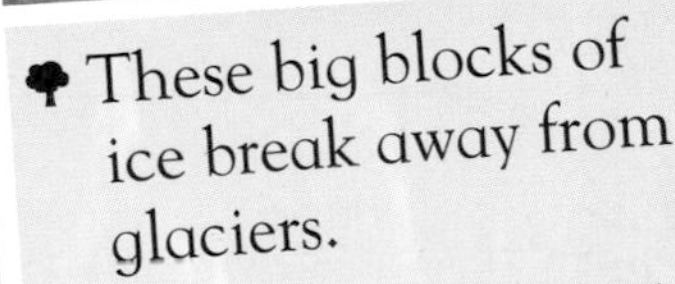

- These big blocks of ice break away from glaciers.
- Two-thirds of this block of ice lies underwater.

See page 29

- They are strangely shaped pillars.
- They form when rain washes away soft rock, exposing hard rock.

See page 23

Index

Acknowledgements

Dorling Kindersley would like to thank:
Dorian Spencer Davies for original illustrations;
Jonathan Brooks for picture library services.

Picture credits:

The publisher would like to thank the following for their kind permission to reproduce their photographs:
a=above; c=centre; b=below; l=left; r=right; t=top

Ardea London Ltd: Graham Robertson 39l. **Bruce Coleman Ltd:** 19t, 40tr, 40br, 44bl; Jules Cowan 22; Pacific Stock 32-3. **Corbis:** 14-5, 21, 23tl, 28-9b, 28-9t, 28r, 29tl, 29r, 32c, 54, 55, 56 boarder; Archivo Iconograofico 26c; Hubert Sadler 16l; Stuart Westmorland 4-5b; Ted Spiegel 17r; Tom Bean 16-7, 52cl, 53br, Hal Beral 53tc, Ralph A. Clevenger 53c, Image Plan 59c, Albert Lleal / Minden Pictures 51cl. **Dorling Kindersley:** Donald Smith - modelmaker 59bc, Dreamstime: Clearviewstock 4-5 (background), Natural History Museum, London 59tl, 59cl (Skull), Pearson Education Ltd 51br, Peter Griffiths - modelmaker 48bl, 59br, Rough Guides 52tr, 52c, 53tl, 58tc, 58c, 58cr, 58crb, 58crb (molten lava), 59tc, 59cra, 59cl, 59cr, Coleman Yuen / Pearson Education Asia Ltd 50br. **Dreamstime.com:** Alexyndr 48tr, Anetlanda 48crb, Shariff Che'Lah 51cra, Doughnuts64 49br, Klikk 59c (Eyjafjallajokull volcano), Laumerle 58tl, Jan Martin Will / Freezingpictures 48c, Daniel Rajszczak 59tr, Timberlakephoto 58bc; **Environmental Images:** Herbert Giradet 42tl. **Gables:** 27r. GeoScience Features. **Fotolia:** Pekka Jaakkola / Luminis 50tc, Silver 58br, uwimages 59ca, Zee 50bc. **Getty Images** joSon / Iconica 19br, Stocktrek Images 59ca (Earth), Telegraph Colour Library 52bc; **Picture Library:** 39cr. **Robert Harding Picture Library:** 10tr, 16cl; Thomas Laird 10c. **Hutchison Library:** 40main. **The Image Bank/ Getty Images:** 1, 4cr, 24c, 37r. **Masterfile UK:** 23br, 36c, 40l, 56c; John Fost 24l. **N.A.S.A.:** 4-5t, 7t. **Natural History Museum:** 17tr. **N.H.P.A.:** Anthon Bannister 34tl; Haroldo Palo Jr. & Alberto Nardi 26-7; Robert Thomas 26bl. **NHPA / Photoshot:** Haroldo Palo Jr 53cl;. **Photolibrary:** Radius Images 50c **Planetary Visions:** 2, 18-9, 40tc, 40cr, 40bl, 41tc. **Powerstock Photolibrary:** 3, 18b. Rafn Hafnfjord: 8-9. **Science Photo Library:** 4-5ca, 12-3, 20l; B. Murton 32-3b, 33tl; Bernhard Edmaier 12l; David Nunuk 23tr, 53bl; ESA 4c G. Brad Lewis 13t; Martin Bond 25, 30. **Still Pictures:** 34c, 34-5, 36bl, 41r, 42cr, 42b, 43, 44cl, 45. **Corbis Stock Market:** 24bc. **Stone/Getty Images:** 11t 14l, 19b, 30-1, 31t. **Telegraph Colour Library/Getty Images:** 5br, 14c, 20c.